BestMasters

Springer awards „BestMasters" to the best master's theses which have been completed at renowned universities in Germany, Austria, and Switzerland.

The studies received highest marks and were recommended for publication by supervisors. They address current issues from various fields of research in natural sciences, psychology, technology, and economics.

The series addresses practitioners as well as scientists and, in particular, offers guidance for early stage researchers.

Sandra Haschke

Electrochemical Water Oxidation at Iron(III) Oxide Electrodes

Controlled Nanostructuring as Key for Enhanced Water Oxidation Efficiency

Foreword by Prof. Dr. Julien Bachmann

Springer Spektrum

Sandra Haschke
Erlangen, Germany

BestMasters
ISBN 978-3-658-09286-3   ISBN 978-3-658-09287-0 (eBook)
DOI 10.1007/978-3-658-09287-0

Library of Congress Control Number: 2015935064

Springer Spektrum

Printed on acid-free paper

Springer Spektrum is a brand of Springer Fachmedien Wiesbaden
Springer Fachmedien Wiesbaden is part of Springer Science+Business Media
(www.springer.com)

Foreword

Saving the planet has become a popular topic among chemistry students and faculty alike. Individual steps of the water splitting reaction have been extensively studied at the interfaces of carbon-based nanostructures, notorious for corroding fast, and of noble metal particles, the cost of which is redhibitory for large-scale applications. In this context, Sandra Haschke's materials choice appears all the more consequent and bold: her Master's thesis is dedicated to iron(III) oxide, 'rust', as a platform for electrochemical water oxidation, the kinetic bottleneck of the overall water splitting.

Indeed, iron(III) oxide provides the advantages of wide availability and low cost, as well as thermodynamic stability, two aspects of crucial practical importance. The impediments associated with iron(III) oxide, however, are of fundamental nature: low electrical conductivity and poor catalytic activity. Recent literature precedents have provided approaches to circumventing these difficulties by doping the solid with various transition metal ions in its bulk or at its surface. Variations of the surface morphology on several length scales (1-1000 nm) have repeatedly been invoked to also influence the overall (photo)electrochemical performance via the geometric effects of surface area and transport path lengths, albeit never directly addressed in experiments.

Sandra tackles that geometric aspect in her work. She establishes the preparation of highly ordered nanoporous iron(III) oxide surfaces the geometry of which is accurately defined and perfectly tunable. She characterizes the chemical and morphological changes that take place upon thermal annealing and electrochemical treatment. She demonstrates that a roughened surface translates into improved water oxidation current via the increased specific surface area. She then systematically investigates how electrocatalytic turnover is affected by elongation of the samples' pores, and identifies a clear optimum value. She finally demonstrates that this optimal geometry corresponds to the experimental situation in which charge transport towards the electrode surface and charge transfer at the surface occur on the same timescale.

Thus, this Master's thesis not only stands out in terms of the breadth of methods utilized in the realms of preparation (porous materials, thin films), characterization (microscopic and spectroscopic techniques), and electrochemical investigation. Rather, it impresses by providing a very complete and coherent picture of the factors controlling electrocatalytic turnover at what is initially considered as an unlikely catalyst candidate, but turns out to be a versatile and well-behaved material when appropriately structured. In fact, over the course of her work, Sandra manages to improve electrocatalytic turnover at iron(III) oxide a thousand-fold. Thereby, she delivers not only a physical-chemical insight of fundamental importance, but also a substantial, quantitative efficiency gain. With this, she puts iron-based materials not quite on par with systems that require much more costly elements, but at least into a range that allows for close comparison — a development of broad significance for chemistry and chemical engineering. In essence, Sandra shows that via clever nanostructuring, rust can be transformed into gold, to some extent...

Prof. Julien Bachmann

Anorganische Chemie, Friedrich-Alexander-Universität Erlangen-Nürnberg

Acknowledgements

I would like to express my gratitude to all the people who helped and supported me in writing this thesis.

First, I would like to thank my supervisor of this project, Prof. Dr. J. Bachmann, who was exceedingly helpful and offered invaluable guidance, support and advice.

Also, I want to take this opportunity to thank my group members for their hearty welcome and several pleasant hours together. Special thanks go to Sebastian and Yanlin for familiarizing me with our equipment and working procedures and for being obliging at all times. Equally, I want to thank Loïc, who always gave me advice and helped me with slightly refreshing my French.

Thanks to Dr. Muhammad Y. Bashouti for measuring XPS with me and analyzing my data.

Last but not least, I express my gratitude to my family, friends and Andy for their support and endless patience.

Sandra Haschke

Table of contents

List of abbreviations

a.u.	arbitrary units
AAO	anodized aluminum oxide
AC	alternating current
ALD	atomic layer deposition
AZO	aluminum doped zinc oxide
B.E.	binding energy
CB	conduction band
CNT	carbon nanotube
Cp	cyclopentadienyl
CV	cyclic voltammetry
CVD	chemical vapour deposition
DC	direct current
EDX	energy dispersive X-ray spectroscopy
EDXA	energy dispersive X-ray analysis
EIS	electrochemical impedance spectroscopy
NHE	normal hydrogen electrode
PEC	photoelectrochemical water splitting
PVC	poly(vinyl chloride)
RF	radio frequency
SEM	scanning electron microscopy
VB	valence band
XPS	X-ray photoelectron spectroscopy

List of Figures

List of Tables

1 Introduction

With the growing challenges of air pollution, the greenhouse effect, related global warming and coincident desire for environmental sustainability, global research has set itself the objective of developing new clean energy technologies. The requirements placed on the transformations of renewable energy sources involve not only high efficiencies, but also the application of inexpensive, abundantly available materials.[1-3]

With the focus on artificial photosynthesis, the latter requirement is not fulfilled by most common catalysts, since the most efficient catalysts base on noble metals. Iron(III) oxide (Fe_2O_3), on the other hand, has the advantages of environmental harmlessness, chemical stability over a wide pH range, and low cost, due to the natural occurrence of iron as the fourth most abundant element in the earth's crust.[4-6] Furthermore, the impediment of the four-electron oxidation of two molecules of H_2O to O_2 is electrochemically achievable near room temperature and neutral pH for the transition metal oxide catalyst .

However, efficient water oxidation with Fe_2O_3 electrodes has not really been achieved yet due to its poor catalytic properties, low electrical conductivity, and the related hindrance of charge transport.[7] These problems have already been addressed in many different ways in recent research.

One strategy is to add impurities such as Ti,[8, 9] Si,[10, 11] or Ni[12] to Fe_2O_3. These dopants lead to an increase in the carrier density and thus to an improvement of the Fe_2O_3 conductivity. In addition to doping the catalyst, the formation of heteronanostructures showed promise. Enhanced charge transport was proven by Lin et al. by means of titanium disilicide ($TiSi_2$) nanonets and conductive aluminum doped zinc oxide (AZO) nanotubes subsequently coated with Fe_2O_3.[13]

Besides these chemical approaches to improve the catalytic properties of the semiconductor material, a fundamentally different and complementary strategy relies on geometry. Indeed, increases in the absolute number of catalytically active surface sites, brought about by controlled nanostructuring of the surface, may also result in the overall improvement in performance parameters. In this context, there have been several reports detailing nanostructuring of Fe_2O_3 including systems such as nanoparticles,[14] nanocolumns,[15] Si-doped nanodendrites[10] and nanoplatelets[16]. For all nanostructured Fe_2O_3 electrodes, electrocatalytic current enhancement was associated with the magnified specific surface area.

Despite this, systematic investigation of the relation between catalytic current and specific surface area has not been widely researched.[3] In this respect, the utilization of a system consisting of straight cylindrical pores with accurately adjustable parameters such as length, diameter and period of the pores is the ideal basis. This model system is created by electrochemical oxidation, also known as "anodization"[17]. Along with the application of porous templates, atomic layer deposition (ALD) is employed, which offers large-scale conformal depositions, easy reproducibility and consequently compatibility with industrial manufacturing.

2 Objectives

As indicated above, Fe_2O_3 is a promising candidate for the oxidative half-reaction of water splitting, which has been the topic of several studies. Nevertheless, distinct deficits in catalytic activity and a systematic relation between catalytic current and surface area are still issues which need to be addressed.

Therefore, this thesis focuses on the electrocatalytic water oxidation with nanostructured Fe_2O_3 electrodes in a phosphate buffer at pH 7. Anodized aluminum oxide (AAO) is chosen as the template for the nanostructures, which is then coated with the catalyst by means of atomic layer deposition (ALD).

The effect of surface area enhancement on the water oxidation efficiency is studied for pore elongation. Here, the optimal system is to be determined in which the maximum catalytic turnover and the related highest overall current density can be achieved.

Linked to system optimization, is the problem of low conductivity of the catalyst surface. For this purpose, the basic nanostructured Fe_2O_3 system is compared with electrodes with underlying conductive layers. Platinum or aluminum doped zinc oxide (AZO), which are deposited by galvanic plating or ALD, respectively, are chosen as underlying layers.

Furthermore, the influence of post-deposition annealing of the Fe_2O_3 surface under O_2 or N_2 atmosphere is investigated with respect to changes in morphology and chemical composition.

In addition to the electrode preparation procedure, methods used are steady-state electrolysis, electrochemical impedance spectroscopy (EIS) for distinction of the different electrochemical processes, scanning electron microscopy (SEM) and X-ray photoelectron spectroscopy (XPS) for structural and elemental surface analysis, respectively.

This work shall therefore serve to optimize the Fe_2O_3 electrode performance and to gain a better understanding of the structural changes and electrochemical reactions taking place with respect to the future application of Fe_2O_3 electrodes for solar water splitting.

3   Fundamentals

3.1   Atomic layer deposition [3, 18, 19]

Conformal coatings in deep pores is preferentially performed by means of ALD. This procedure is basically derived from chemical vapor deposition (CVD), with the peculiarity that the reaction between the reactants occurs at the sample surface and not in the gas phase. In ALD, the sample surface is exposed in an alternating manner to two gaseous precursors. At each exposure, the incoming precursor molecules are chemisorbed at the surface until one monolayer is formed. The adsorption then stops regardless of the amount of precursor entering the reaction chamber. The unreacted molecules and gaseous by-products are subsequently removed by evacuation before the second precursor is let into the chamber and quantitatively reacts with the former in a similar manner. After the second evacuation, the procedure can be repeated until the desired thickness is reached. This approach therefore contains well-defined surface chemistry resulting in self-limiting growth and precisely controllable layer thicknesses. Accurate Fe_2O_3 layers, which are essential for this work, are obtained by the well-established oxidation of ferrocene by ozone at 200 °C.

3.2   Water splitting at semiconductor electrodes [20-26]

Fujishima and Honda[27] performed pioneering work concerning the photoelectro-chemical (PEC) water splitting at semiconductor electrodes. Since their first observation of water photolysis at titanium dioxide (TiO_2) electrodes, significant research efforts have been devoted to identifying alternative semiconductors. The two most important requirements on the materials are a valence band (VB) energy more positive than the standard potential of the redox couple H_2O/O_2 and a conduction band (CB) energy more negative than the H_2O/H_2 level. Furthermore, a relatively small band gap is advantageous for harvesting visible solar energy. These demands on the band positions are illustrated for TiO_2 and three other potential candidates in Figure 1. Especially the last requirement, is only partly fulfilled for TiO_2 which just absorbs the ultraviolet part of the sun emission due to its band gap of 3.0 - 3.2 eV. The transition metal oxide semiconductor Fe_2O_3, on the other hand, has an ideal band gap of 2.0 - 2.2 eV, however, its CB edge is lower than the H_2O/H_2 level. The spontaneous electron transfer for water reduction is thus hindered, whereby an additional applied potential is necessary for this reaction. The application of Fe_2O_3 as a photoanode for water oxidation on the contrary is quite favorable, since the VB is located below the H_2O/O_2 level with $E_{VB} = -6.98$ eV.

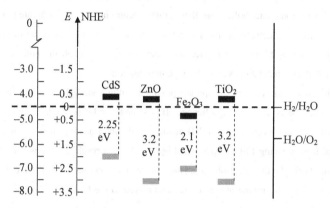

Figure 1. Band positions of several semiconductors in contact with an aqueous electrolyte at pH 1. The energy scale is indicated in electron volts using either the normal hydrogen electrode (NHE) or the vacuum level as a reference. On the right side the standard potentials of the redox couples H_2O/H_2 and H_2O/O_2 are presented. The valence band and conduction band are illustrated with grey and black bars. Values are taken from a review on photoelectrochemical cells.[20]

In the following, the band model of a Fe_2O_3 electrode, which will come in contact with an aqueous electrolyte, is considered more precisely (see Figure 2). In the initial state, the Fermi level of the n-type semiconductor $E_{F,SC}$ is closer to the CB due to its excess in electrons and the Fermi level of the electrolyte $E_{F,EL}$ (defined by the electrochemical potential of the H_2O/O_2 redox couple) is below $E_{F,SC}$ (Figure 2a). The subsequent electrolyte contact leads to an electrochemical equilibrium at the interface associated with the adjustment of a common E_F. The mobile nature of both the charge carriers and the counter-ions in the electrolyte causes the potential of the solution to remain essentially constant and homogeneous, so that the common E_F will be near $E_{F,EL}$. This equilibration, which corresponds to an exchange of electrons across the interface, is accompanied by the formation of a region on either side of the interface in which the charge distribution differs from the bulk material, the so-called space-charge layer. In this respect, a depletion layer is formed in Fe_2O_3 due to the electrons that have left the solid for the solution, leaving behind a positive excess charge. The resulting potential difference in the solid is manifested by bands bending upwards in the vicinity of the interface (Figure 2b).

If this electrode-electrolyte system is now exposed to light, electron-hole pairs are generated, whereby free electrons are generated in the CB and holes are left behind in the VB. This process is accompanied by the splitting of $E_{F,SC}$ into quasi Fermi levels $E_{F,n}$ and $E_{F,p}$ (Figure 2c). Electrons and holes can then either recombine or are separated, which is favorable in terms of electrical energy production. In the latter case, holes are driven to the surface and scavenged by the electrolyte, and electrons move through the bulk towards the charge collector. This results in overall charge compensation.

For this work, the case of an applied potential without sunlight exposure at a solid-electrolyte interface is of more significance. If a strong anodic potential is applied, $E_{F,SC}$ moves below $E_{F,EL}$ and even below the VB position at the interface which is accompanied by an enhanced band bending further upwards (Figure 2d). As a result, electrons flow from the VB into the bulk. If the electron concentration is now below the intrinsic level, the semiconductor is locally p-type at the surface and n-type in the bulk. This layer is known as inversion layer.

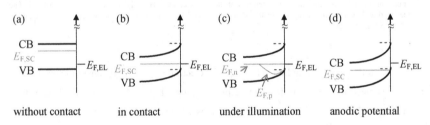

Figure 2. Band model of a n-type semiconductor electrode (solid to the left and electrolyte phase to the right of arrow). (a) semiconductor and electrolyte independent of each other, (b) band bending due to electrolyte contact, (c) band splitting due to illumination, (d) adjustment of Fermi level $E_{F,SC}$ by anodic potential application; $E_{F,SC}$ and $E_{F,EL}$ Fermi levels of semiconductor and electrolyte, VB and CB valence and conduction band.

3.3 Electrochemical impedance spectroscopy [28-30]

The electrochemical behavior of systems consisting of a solid-liquid interface in which the overall current is determined by a number of strongly coupled processes, each proceeding at a different rate, can be characterized by EIS. This technique investigates the system response to an applied sinusoidal AC potential ΔE, superimposed on the static potential E. The amplitude of ΔE is small, typically 10 mV, and the measurements are carried out at a range of frequencies ω typically between MHz and mHz, whereby the resulting oscillating current $I(\omega)$, phase-shifted by ϕ, is recorded, from which an impedance can be calculated by

$Z(\omega) = E/I = Z_0(\cos\phi + i\sin\phi)$. Z includes a real (Z_{real}) and an imaginary (Z_{imag}) component. The gained data can be represented either in a "Bode plot" ($|Z(\omega)|$ and ϕ vs. $\log\omega$) or in a "Nyquist plot", where $-Z_{imag}$ vs. Z_{real} is plotted. The shape of the plot suggests an equivalent circuit which may give rise to a similar electrical response. Fitting the values of the electrical elements to the EIS data provides insight into changes in solid structure and electrochemical reactions (mass transport, charge transfer, recombination, *etc.*) taking place.

4 Methods and materials

4.1 Materials

Ethanol, perchloric acid (70%), chromium (VI) oxide, phosphoric acid, copper (II) chloride dihydrate, hydrochloric acid (30%), nickel (II) chloride hexahydrate, nickel (II) sulfate hexahydrate, boric acid, sodium hydroxide, sulfuric acid (96%), hexachloroplatinic acid hexahydrate (30 - 40% Pt), potassium dihydrogen phosphate, diethylzinc, trimethylaluminum and ferrocene (purity = 99%) were purchased from Sigma Aldrich, Alfa Aesar, ABCR and VWR and used as received. Water was purified in a Millipore Direct-Q System for the application in electrolytes. Aluminum plates (99.99%) were supplied by SmartMembranes. For planar substrates, Si (100) wafers covered with an oxide layer were purchased from Silicon Materials Inc.

4.2 Preparation of Fe_2O_3 electrodes

Fe_2O_3 electrodes were produced in a multistep process delineated in Scheme 1. The preparation started with a two-step anodization procedure of aluminum in order to gain aluminum oxide membranes utilized as templates for nanostructered electrodes. Anodization was performed in home-made two-electrode cells consisting of a PVC beaker with four circular openings at the bottom, under which aluminum plates of 2 cm diameter were held between an O-ring and a thick copper plate operating as an electrical contact. The PVC beaker contained the electrolyte and was provided with a lid equipped with a mechanical stirrer and silver wire mesh as counter-electrode, whereas the copper plate was in contact with a cold plate connected to a closed-circuit cooler by Haake. The whole setup was thermally insulated laterally. At first, the aluminum plates were electropolished in a cooled perchloric acid/ethanol solution (1:3 v/v $HClO_4$/EtOH) for 5 min under +20 V. They were then rinsed, cooled and anodized under +195 V for 23 h at 0 °C in 1 wt.% H_3PO_4. Aluminum oxide was removed by chromic acid solution (0.18 M CrO_3 in 6 wt.% H_3PO_4) for 23 h at 45 °C. Subsequently, the second anodization was performed for 3, 4, 6 and 8 h at 0 °C in 1 wt.% H_3PO_4 in order to vary the pore length. The next steps involved detaching the aluminum on the backside of the Al_2O_3 pores with 0.7 M $CuCl_2$ solution in 10% HCl, removing the samples from the electrochemical cell, eliminating the barrier oxide closing the pores and simultaneous isotropic pore widening in 5 wt.% H_3PO_4 at 45 °C for 28 - 30 min. A small proportion of samples received a unilateral treatment for 52 min in the cell.

The anodic alumina templates were then coated in a sputter coater by Torr International Inc. with 80 ± 16 nm of gold in the DC mode (40 mA). For subsequent electroplating of the thicker nickel layer, the samples were again mounted into the electrochemical cells with only one circular opening, which was filled with diluted standard Watts electrolyte (0.285 M $NiSO_4 \cdot 6H_2O$, 47.5 mM $NiCl_2 \cdot 6H_2O$, 0.5 M H_3BO_3 at pH = 3.0) and provided with a platinum mesh as counter-electrode. Nickel was electrodeposited at −2 V ($I = 0.10$ mA for $D_{sample} \sim 2$ cm) for 80 min.

The last preparation step included the atomic layer deposition (ALD) of Fe_2O_3 in a home-built hot-wall reactor operating with N_2 as carrier gas. The reactor was fitted with a chemically resistant diaphragm pump MV10C from Vacuubrand. ALD was performed at 200 °C in the chamber with ferrocene, which was kept in a stainless steel bottle maintained at 80 °C and with O_3 from an OzoneLabs generator. Pulse, exposure and purge times were 2, 50, 60 s for ferrocene and 0.2, 50, 60 s for O_3. 900 cycles were performed in order to obtain approximately 10 nm of Fe_2O_3.

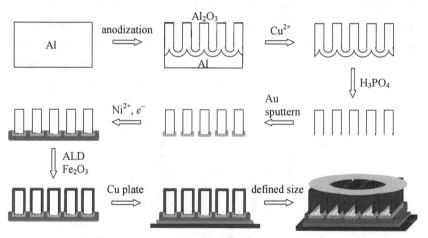

Scheme 1. Preparation of nanostructured electrodes of defined pore length, diameter and period as well as of defined thickness of the Fe_2O_3 layer. The circle represents a kapton mask.

4.3   Production of modified Fe_2O_3 electrodes

4.3.1 Annealed catalyst surfaces

Some of the resulting samples were annealed in a high-temperature furnace from the company Nabertherm under O_2 or N_2 atmosphere. In this step, the samples were heated up to 400 °C over 12 h, kept at this temperature for 4 h and cooled down to room temperature over 12 h.

4.3.2 Electrodes with underlying platinum layers

Underlying platinum layers were obtained by galvanic methods. To this goal, unilaterally opened porous templates were sputter coated with 500 ± 100 nm of a copper-gold mixture as an electrical contact. The samples were then placed with their electrical contact onto small copper plates and adjusted into electrochemical cells with only one circular opening exposing the other side to the platinum electrolyte (4 mM $H_2PtCl_6 \cdot 6H_2O$ solution adjusted to pH = 5 with 1 M H_2SO_4). The galvanic plating was performed at room temperature on a Gamry Reference 600 potentiostat in a three-electrode configuration (see Figure 3). Therefore, the cell was fitted with a platinum mesh auxillary electrode and an aqueous Ag/AgCl/KCl(sat) reference electrode from BASi (Bioanalytical Systems, Inc.). Nanotubes were grown for 10.5 min with alternating potentials of –400 mV for 2 min and 0 V for 4 min. Subsequently, Fe_2O_3 was deposited with 800 cycles of ALD as described above.

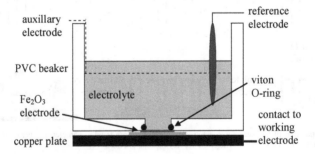

Figure 3. Delineation of the electrochemical three-electrode setup.

4.3.3 Electrodes with underlying AZO layers

The modification with aluminum doped zinc oxide (AZO) was approached by ALD on unilaterally opened porous templates sputter coated with 500 ± 100 nm of a copper-gold mixture as backside contact. ALD of AZO was performed in a GEMSTAR-6 [TM] Benchtop ALD reactor by Arradiance[31] operating with N_2 as carrier gas. The reactor was provided with a pump protection exhaust abatement system and a rotary vane pump. Diethyl zinc and deionized water were used as the precursor reactants to deposit the ZnO film and trimethylaluminum and H_2O to deposit Al_2O_3. The chamber temperature was set to 120 °C. Pulse, exposure and purge times of DEZ, TMA and H_2O were 0.2, 40 and 60 s. The AZO films were deposited with an TMA to DEZ ratio of 10%. The doping concentration was achieved by 9 cycles of DEZ/H_2O and 1 cycle of TMA/H_2O. This "supercycle" was repeated 5 times in order to gain approximately 10 nm of AZO after a total number of 100 cycles. One piece of a titanium tube was used to control the presence of aluminum in the deposit. AZO films were partly annealed for 2 h at 450 °C in N_2 atmosphere for subsequent resistivity measurements. At last, the AZO film was coated with 800 cycles of Fe_2O_3 in the home-built reactor.

4.4 Preparation of planar electrodes

For the preparation of planar electrodes, Si (100) wafers were used as substrates. 5 ± 1 nm of aluminum were sputter coated in the RF mode (25 W) as an adhesion layer followed by 80 ± 1 nm of gold in the DC mode. ALD of Fe_2O_3 was performed as described above only differing in the pulse, exposure and purge times of 2, 15, 30 s and 0.2, 15, 30 s for $FeCp_2$ and O_3.

4.5 Instrumental methods

The thickness of the deposited Fe_2O_3 layer was determined for a Si (100) wafer as reference with a spectroscopic ellipsometer EL X-02 P Spec by DRE-Dr.Riss Ellipsometerbau GmbH. The ellipsometer was adjusted with a halogen lamp (35 W, white light) and a monochromator (400 nm - 1000 nm) prior to the sample. For each measurement 50 data points were recorded for wavelengths between 400 and 1000 nm and for an angle of incidence of 70°. The data, the orientation Θ and ellipticity ε, were then fitted to an optical model which contained the thickness and optical constants of the investigated layer (see Figure S 1).

Scanning electron microscopy (SEM) and energy dispersive X-ray spectroscopy (EDX) were obtained on a JEOL JSM 6400 PC at 10 - 20 kV and implemented with a LaB_6 cathode. Cross-sections of the membranes were studied at the edges of broken samples.

X-ray photoelectron spectroscopy (XPS) was performed in a home-made spectrometer. Core level and valence band photoelectron spectra were recorded upon excitation with monochromatic Al K radiation (1487 eV) and collected, at a take-off angle of 35°, by a hemispherical analyzer with an overall resolution adjustable between 0.8 and 1.2 eV. The surveys were conducted in various ranges of electron energies including the overall binding energy survey (0 to 1200 eV), with the addition of individual spectra for Si 2p (95 to 110 eV), C 1s (282 to 287 eV), N 1s (390 to 410 eV), P 2p (130 to 145 eV), O 1s (520 to 550 eV) and Fe 2p (700 to 730 eV) which were monitored more accurately in a defined number of scans. Scan times of up to ~ 30 min were employed for all data collections with high resolution. All spectra were taken at room temperature in a UHV chamber under about 10^{-10} Torr pressure. The resulting XPS spectra were analyzed by spectral decomposition using the XPS peak Sigma Probe Advantage software.

4.6   Electrochemical studies

The nanostructured electrodes were laser-cut with a GCC LaserPro into pieces, adhered with the nickel contact onto small copper plates with the aid of a conductive silver adhesive and mounted from the top with kapton tape which exposed a hole of defined diameter $d = 1 - 2$ mm (Figure 4a and b).

For investigation of planar samples, the coated wafer was placed on a small copper plate, whereby the electrical contact between the Fe_2O_3 surface and the copper plate was established by the silver adhesive frame at the edge of the wafer. For direct comparison with the nanostructured electrodes, the Fe_2O_3 surface was covered with kapton tape holding commensurate holes (Figure 4c and d). The small copper plate was then adjusted into an electrochemical cell, as described for electroplating of platinum, exposing the defined sample area to the phosphate electrolyte prepared from 0.1 M KH_2PO_4 adjusted to pH 7 (internal resistance $R = 30 - 40\ \Omega$). All electrochemical measurements were performed at room temperature on a Gamry Reference 600 and Interface 1000 potentiostat in the standard three-electrode configuration (see Figure 3). The standard redox potential of the reference electrode is shifted by +0.20 V relative to the normal hydrogen electrode (NHE).

Figure 4. Photographs of (a) a nanostructured sample, (b) a prepared Fe_2O_3 electrode on a copper plate mounted with kapton tape with a hole of $d = 2.0$ mm, (c) a coated Si wafer contacted to a copper plate by silver adhesive, (d) a coated Si wafer contacted to a copper plate and covered with kapton tape with a hole of $d = 2.0$ mm. The photographs display the homogeneity of the samples, and the accurately defined circular opening exposed for electrochemical investigation.

For each sample, internal resistance measurements, cyclic voltammetry (CV) and electrochemical impedance spectroscopy (EIS) were carried out without stirring. Cyclic voltammograms were measured at a scan rate of 50 mV s^{-1}. Impedance data were collected using 10 mV amplitude perturbation between 10,000 and 0.02 Hz and analyzed with Gamry Echem Analyst in order to generate the fits presented. Planar and nanoporous electrodes were electrochemically pre-treated over 0.5 - 3 h with an applied potential of +1.3 V, before bulk electrolysis was performed at +1.1 V for at least 16 h under stirring.

Resistivity measurements of AZO coated Al_2O_3 templates were carried out with an EC epilson potentiostat by BASi. The exposed sample area of the prepared nanostructured electrodes was coated in the sputter coater with 5 nm of aluminum followed by 5 nm of gold. This layer was subsequently contacted with a thin copper wire with the aid of silver adhesive. The electrodes were connected from both sides (copper plate and wire) and I-V curves were recorded between +0.5 and −0.5 V. These data were used to calculate specific volume resistances ρ ($\rho = R \cdot \frac{A}{l}$).

5 Results and discussion

5.1 Chemical and structural properties of nanostructured Fe_2O_3 electrodes

Nanostructured Fe_2O_3 electrodes were gradually built up starting with the preparation of Al_2O_3 membranes as templates. Upon adequate cooling ($T = 0\ °C$), the two-step anodic oxidation at +195 V in 1% H_3PO_4 yielded Al_2O_3 layers involving hexagonal arrays of pores, whereby the length was set by the duration of the second anodization. In this case, 3, 4, 6 and 8 hours led to pore lengths varying between 7 and 23 μm. The exact data are listed in Table 1.

Table 1. Results of structural properties of self-ordered porous alumina analyzed from SEM images; t = anodization time, L = pore length, D = pore diameter of coated membranes.

t / h	L / μm	D / nm
3	6.8 ± 0.3	279 ± 36
4	9.5 ± 0.1	305 ± 27
6	14.0 ± 0.3	332 ± 29
8	22.8 ± 0.1	326 ± 23

The lateral period D_p amounted to 500 ± 30 nm, which corresponds to values published by Nielsch et al.[32] After elimination of the underlying aluminum substrate, the backside Al_2O_3 barrier layer, still closing the pore extremities, was removed and simultaneous pore widening was performed via bilateral acidic treatment for 30 min for membranes anodized from 4 to 8 hours. Since shorter Al_2O_3 pores were less stable without their barrier layer, the duration of acid treatment had to be reduced to 28 min for 7 μm membranes. The electrical nickel contact obtained by means of electrodeposition exhibited thicknesses varying between 0.8 and 1.3 μm, whereby the metal primarily grew in the pore extremities as visible in Figure 5a at the left end of the 14 μm long membrane.

In the last step of membrane preparation, the array of pores was coated with Fe_2O_3 by means of atomic layer deposition (ALD). Ensuing from the average deposition rate of 0.1 Å/cycle, 900 cycles were carried out in order to deposit approximately 10 nm of catalyst. The layer thickness was determined by means of ellipsometry (see Figure S 1). Figure 5 illustrates the cross-section (a) and surface (b) of a typical sample achieved by this procedure.

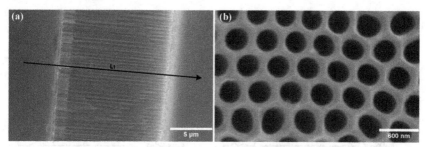

Figure 5. SEM micrographs of (a) the cross-section including the line defined for EDX measurements of Figure 6b, (b) the surface of a 14 μm membrane coated with ~ 10 nm of Fe_2O_3. The micrographs demonstrate the hexagonal order of parallel, cylindrical pores, as well as the homogeneity of their length and diameter.

The SEM images of the coated samples provide the pore diameters D reduced to their final values of 280 to 330 nm for membranes of differing length (see Table 1). At this point the question arises whether the difference in pore diameter allows for a direct comparison of the prepared systems. Therefore, the microscopic area A_{micro} is calculated for a defined macroscopic area of 0.03 cm^2 for all membranes and linear regression is applied subsequently to these values (see Table S 1). Here, A_{micro} consists of the surface area of all pores which are located in the circular area A_{macro} set by the kapton mask. According to the coefficient of determination of $R^2 = 0.9709$ and the diameter deviation of 10 % compared to length deviations of 100 -1000 %, a reliable comparison is possible.

The coating homogeneity is verified with energy dispersive X-ray analysis (EDXA). In the EDX spectrum of the sample depicted in Figure 6a, the most conspicuous peaks are assigned to the elements Al, O, Fe and Ni as expected. The element P is equally present as traces of phosphoric acid can still exist in the nanoporous Al_2O_3. In addition, completely coated pores can be proven by further EDX measurements of the membrane cross-section recorded for 50 points along a defined line indicated by the black arrow in Figure 5a. The intensities obtained for the elements Al, Fe and Ni are shown in Figure 6b. The large intensity difference between Al and Fe is obvious at first sight, which was expected since Fe_2O_3 is only present as a thin coating in the pores of the Al_2O_3 matrix. Nickel is in direct contact with the catalyst layer mostly present inside the pores, which was already evident from the SEM micrograph. The most important aspect is that the Fe intensity is measured with a fairly constant value throughout the whole membrane. The presence of Fe_2O_3 along the whole length of each pore ensures that every point of the surface exposed to the electrolyte is electrically contacted.

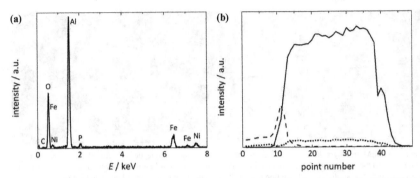

Figure 6. (a) EDX spectrum of the cross-section of the 14 μm membrane visualized in Figure 5a, (b) EDX data of Al (solid line), Ni (dashed line) and Fe (dotted line) for 50 points recorded along the black arrow in Figure 5a. The element Fe is present in homogeneous concentration throughout the depth of the pores, whereas the Ni contact exists only on one side of the sample.

For subsequent electrochemical studies, coated membranes were cut into pieces, adhered with the nickel contact onto a small copper plate and confined to defined macroscopic area of $3.1 \cdot 10^{-2}$, $1.8 \cdot 10^{-2}$ or $7.9 \cdot 10^{-3}$ cm^2 by means of a circular kapton mask. This work is mostly focused on nanoporous electrodes, whereas planar electrodes, based on Si wafers coated with Al/Au (5 ± 1 nm/80 ± 16 nm) and Fe$_2$O$_3$ (~ 10 nm), serve as reference and partly as simplified system for the investigation of chemical and structural changes.

5.2 Electrochemical activity of Fe₂O₃ electrodes

Electrochemical experiments were carried out in electrochemical cells in which the Fe$_2$O$_3$ electrodes face the phosphate electrolyte on a defined area. The macroscopic area A_{macro}, which is defined by the circular area of the kapton mask, is used for the determination of the effective current density J presented in the graphs. In one case, the effective current density is referred to the microscopic area A_{micro} and is thus denoted as J_{micro}. A_{micro} consists of the surface area of all pores located in the circular area A_{macro}. For all nanostructured electrodes a series of measurements was performed in the following order:

(1) cyclic voltammetry

(2) electrochemical pre-treatment over 0.5 - 3 h at +1.3 V

(3) cyclic voltammetry

(4) bulk electrolysis at +1.1 V for 16 h

(5) cyclic voltammetry

(6) electrochemical impedance spectroscopy.

Prolonged water electrolysis is recorded at lower potentials $E = +1.1$ V which corresponds to an overpotential η of 0.49 V vs. Ag/AgCl, $\eta = \Delta E - E^{\circ}(O_2/H_2O) + 0.060$ V $\cdot$ pH (where $E^{\circ} = 1.23$ V and ΔE : applied potential vs. Ag/AgCl).

5.2.1 Electrochemical pre-treatment

Starting with the electrochemical preparatory treatment of the electrodes, it is useful to clarify the necessity of this procedural step. Referring to earlier studies on nanoporous Fe$_2$O$_3$ electrodes[3], bulk electrolyses at +1.3 V were implemented on the first prepared samples, whereby a strong current increase over the first few hours was detectable before reaching a constant value (see Figure 7). It is obvious that the current enhancement can be limited to approximately the first three hours. On the basis of this observation, the effect of different pre-treatment durations of 0.5 to 3 hours is studied for 23 μm membranes in respect of the effective current densities achieved after 16 h of bulk electrolysis. For each pre-treatment two measurements are performed. Since the electrodes under investigation are differing in pore diameter due to uni- and bilateral removal of barrier oxide at the Al$_2$O$_3$ templates, microscopic current densities J_{micro} of the corresponding microscopic areas are summarized in Table 2. The obtained values are all in the same order of magnitude, whereby 1 h and 2 h of

pre-treatment seem to be the most promising times. In the following, electrochemical
preparatory treatments were set to one or two hours at +1.3 V depending on the experiment.

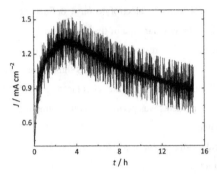

Figure 7. Effective current density J recorded at +1.3 V over 15 h at structured Fe_2O_3 electrodes coated with
~ 10 nm of Fe_2O_3; macroscopic area $A_{macro} = 3.1 \cdot 10^{-2}$ cm^2; $L = 22$ μm and $D = 188 \pm 9$ nm.

Table 2. Microscopic current densities J_{micro} recorded for differing pre-treatment times t ($E = +1.3$ V) after 16 h
of bulk electrolysis at +1.1 V at structured Fe_2O_3 electrodes; $A_{macro} = 3.1 \cdot 10^{-2}$ cm^2, $L = 23$ μm, standard
deviation of all values is $\sigma = 0.002$ μA cm^{-2}.

t / h		0.5	1	2	3
J_{micro} / μA cm^{-2}	(1)	0.09	0.16	0.14	0.10
	(2)	0.11	0.12	0.07	0.11
Ø / μA cm^{-2}		0.10 ± 0.01	0.14 ± 0.03	0.10 ± 0.05	0.10 ± 0.01

In the next step, the origin of this activity increase is of interest. In order to attain a
better understanding of the chemical and/or structural changes associated with electrolysis,
X-ray photoelectron spectroscopy (XPS) and electrochemical impedance spectroscopy (EIS)
were performed at planar reference samples.

Beginning with XPS investigations of untreated Fe_2O_3 coated Si wafers in comparison
to wafers which received 2 h of pre-treatment at +1.3 V, the samples were initially scanned
from 0 to 1200 eV in order to obtain an overview of the measurable elements. Such an XPS
spectrum of an ALD-prepared sample is visualized in Figure 8a presenting the most
conspicuous peaks corresponding to the elements O, Fe and C. For detailed analyses, high-
resolution scans of the individual peak regions were performed, followed by deconvolution
and multiplet fitting of the recorded data (see Figure 8b) to gain the overall peak shapes,

binding energies *B.E.*, intensities and element ratios listed in Table S 3, Table S 4 and
Table 3.

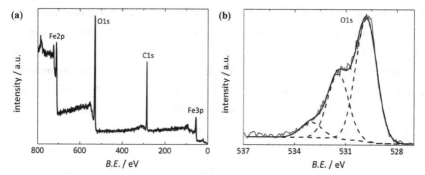

Figure 8. (a) XPS spectrum of a planar ALD-prepared Fe_2O_3 electrode, (b) deconvolution of the O 1s peak region; raw data (grey line), individual fits (dashed black line) and the corresponding total fit (solid black line).

Table 3. Arbitrary intensity ratios determined for pairs of XPS peaks of planar untreated and electrolyzed (2 h, +1.3 V) Fe_2O_3 electrodes; calculated from intensities listed in Table S 3 and Table S 4. Fe/O_{Fe} ratio of the electrolyzed sample is computed from the total intensity of Fe and the intensity of the first O 1s peak.

ratio	untreated	untreated + electrolysis
Fe/O_{Fe}	0.78	1.07
P/Fe	0	0.02
Si/Fe	0	0.25

The spectrum of the Fe 2p region (Figure 9a) and the binding energies associated with it clearly demonstrate that the iron oxidation state does not change upon electrolysis, as no shift in the *B.E.* position is observable. Furthermore, the $2p_{3/2}$ and $2p_{1/2}$ can definitely be attributed to Fe^{3+} in Fe_2O_3.[33] Electrolysis, however, results in a serious variation of the O 1s region (see Figure 9b). The peak at ~ 530 eV, attributable to the Fe_2O_3 oxide, loses intensity upon electrolysis.[33] Simultaneously, a new peak arises at 532.2 eV, which can be assigned to the oxide of SiO_2, in accordance with the appearance of Si signals in the spectrum.[34] These changes might be indicative of a transition from a smooth, continuous catalyst layer to an irregular one that has lost continuity, since SiO_2 is detectable. Consistent with this interpretation, small cracks on the surface of this sample are apparent on the microscopic scale, which naturally lead to SiO_2 signals. Additionally and more importantly, is the Fe to

O_{Fe} ratio which is now calculated for both measurements. Here the intensity of the O 1s peak at 532 eV (belonging to SiO_2) is not taken into account for the electrolyzed electrode. The outcome is an increase in the iron amount. This raise and the unchanged oxidation state of Fe^{3+} lead to the conclusion that electrolysis causes a decrease in the hydration state. If the XPS Fe/O_{Fe} intensity ratio of 1.07 obtained after treatment corresponds to the stoichiometric ratio 0.67 of Fe_2O_3, then the intensity ratio 0.78 characterizing the untreated sample amounts to a stoichiometry near 0.5, compatible with a formulation FeO(OH). One last important aspect has to be mentioned, which is the occurrence of P 2p peaks after electrolysis. The *B.E.* of 133.3 eV is ascribable to phosphate anions,[35] which likely incorporate into the oxide structure and potentially enhance the catalyst activity.[36, 37]

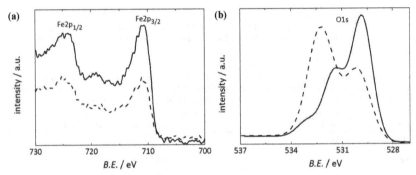

Figure 9. XPS spectra of (a) the Fe 2p and (b) O 1s peak region. Comparison is drawn between an untreated electrode (solid line) and one that has been electrolyzed at +1.3 V for 2 h (dashed line).

The variations in electrochemical characteristics associated with the chemical changes demonstrated by XPS are characterized by EIS. In this context, EIS was carried out on planar electrodes after ALD preparation, 2 h of pre-treatment and after 16 h of bulk electrolysis. Data were collected for frequencies between 20 mHz and 100 kHz at +1.1 V *vs.* Ag/AgCl. Nyquist plots of the corresponding data are displayed in Figure 10 including the fits resulting from the electrical circuit. The equivalent circuit elements include a solution resistance R_s between the reference and working electrode, a charge transfer resistance R_{ct} associated with the electrochemical reaction (a rough estimation of R_{ct} is given by the diameter of the semicircle) and a constant phase element. A constant phase element represents an unideal capacitor with a frequency dependence of the capacitance, the impedance of which is then expressed as $Z = Q^{-1} \cdot (i\omega)^{-\alpha}$, with $\alpha < 1$. This behavior of an unideal capacitor is generally attributed to disturbed surface reactivity, surface inhomogeneities and to current and potential distributions associated with electrode geometry.[30, 38, 39] Since surface irregularities and breaks are known, and simple double layer capacitors C_{dl} exhibit little correlation in the high frequency region, Q is considered more appropriate. The use of this circuit seems further to be promising as the same circuit composition was published recently for hematite thin films.[40]

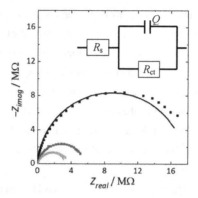

Figure 10. Nyquist plots for EIS data measured at +1.1 V for one planar Fe_2O_3 electrode as deposited (black points), after 2 h of pre-treatment (grey points) and after 16 h of bulk electrolysis (light grey points) including fits (lines) created from equivalent electrical circuit, used to conceptualize Fe_2O_3-liquid interfaces of flat electrodes; R_s = solution resistance, R_{ct} = charge transfer resistance, Q = constant phase impedance and α = exponent representing the capacitor ideality; $A_{macro} = 1.3 \cdot 10^{-1}$ cm^2.

Table 4. Fitting parameters for the EIS data visualized in Figure 10 for one planar Fe_2O_3 electrode as deposited, after 2 h of pre-treatment and after 16 h of bulk electrolysis; R_s = solution resistance, R_{ct} = charge transfer resistance, Q = constant phase impedance (σ is in range of nF s$^{\alpha-1}$) and α = exponent representing the capacitor ideality.

circuit element	R_s / Ω	R_{ct} / MΩ	Q / μF s$^{\alpha-1}$	α
as deposited	284 ± 2	17.8 ± 0.2	0.12	0.96
after 2 h of 1.3 V	259 ± 2	5.7 ± 0.06	0.32	0.88
after 16 h of 1.1 V	236 ± 2	3.3 ± 0.03	0.35	0.86

Independent of the fits, Nyquist plots for all three measurements are simple semicircles representing electron transfer-limited processes,[41] which is expected for planar electrodes. The fitting parameters gained for the measurements are listed in Table 4. R_s amounts to 240 - 280 Ω, which corresponds to internal resistances measured before each EIS. Furthermore, the freshly prepared electrode behaves almost as an ideal capacitor ($\alpha = 0.96$), whereby after electrochemical pre-treatment and bulk electrolysis stronger deviations are visible ($\alpha = 0.88$ and 0.86). This is in line with XPS results, where SiO_2 peaks appeared after 2 h of electrolysis. Concomitantly, the constant phase impedance Q increases by almost a factor of three, which is attributable to surface expansion at the solid-liquid interface. These changes in α and Q, associated with electrolysis, confirm the transition to an irregular, and increased catalyst surface. Simultaneously, this should be linked to a diminution in R_{ct} by at least a factor of three. R_{ct} indeed reduces by more than one-third upon electrochemical pre-treatment and bulk electrolysis. Thus, electrolysis enhances the catalytic activity of Fe_2O_3 predominantly due to crystallization into a rough surface with an increased absolute number of catalytically active surface sites. Additionally, chemical changes in the Fe_2O_3 surface may contribute to a minor extent to the catalyst activity. In this respect, the adsorption or incorporation of phosphate anions into the oxide structure, which is proven by XPS analysis, might be an issue. Independently of the cause, pre-electrolysis at $+1.3$ V is significant for the catalyst performance, since subsequent prolonged bulk electrolysis at $+1.1$ V results only in a small resistance reduction.

To conclude, an important current increase is observed within the first few hours of electrolysis at $+1.3$ V. This effect is in the following specifically induced by pre-treating the catalyst surfaces at $+1.3$ V for one or two hours in order to achieve constant values earlier and to enhance the catalyst activity in bulk electrolysis at a lower overpotential η.

5.2.2 Bulk electrolysis of water at nanostructured Fe_2O_3 electrodes

In the following, the performance of the nanoporous electrodes in prolonged water oxidation is investigated. To this goal, electrochemical experiments are carried out as listed at the beginning of chapter 5.2 (CV, pre-treatment, CV, bulk electrolysis, CV and EIS). Results are presented for one electrode prepared from an Al_2O_3 template with 14 μm long pores covered with 10 nm of Fe_2O_3 as an example for electrochemical measurements at nanostructured electrodes. The catalyst faces the aqueous phosphate buffer on a macroscopic area of 0.03 cm².

As in the case of the planar samples, the preparatory treatment over 1 h is associated with a distinct current increase (see Figure S 2a). For subsequent bulk electrolysis over 16 h at reduced overpotential, a stable current density of $J = 19.4$ μA cm^{-2} is achieved (Figure S 2b). It is noteworthy that the reduction in η from 0.69 V to 0.49 V is associated with a decrease in anodic current by at least one order of magnitude. Current enhancement is further observable in the cyclic voltammograms presented in Figure 11, which were recorded for the same electrode immediately after ALD, after pre-treatment and prolonged electrolysis. For each I-V curve, the current corresponding to the maximum applied potential of +1.3 V is definitely larger than the previous one. Moreover, all curves exhibit slightly hysteretic behavior which can presumably be attributed to capacitive and pseudo-capacitive effects, which are in focus of current research.[2, 5, 42]

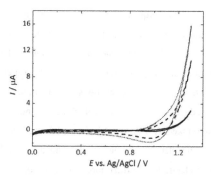

Figure 11. Cyclic voltammogram of a nanostructured electrode as deposited (solid line), after 1 h of pre-treatment at +1.3 V (dashed line) and after 16 h of bulk electrolysis at +1.1 V (dotted line) recorded in phosphate buffer. Scan rate : 50 mV s^{-1}, $A_{macro} = 3.1 \cdot 10^{-2}$ cm², $L = 14$ μm and $D = 332 \pm 29$ nm.

The nanoporous electrodes were mostly stable upon prolonged electrolysis, since no significant differences were observed at the microscopic scale (see Figure 12a and b). A few limited exceptions exist which exhibited cracks of macroscopic size. Moreover, the elemental composition of the surface before and after water oxidation is basically identical according to EDX analyses (see Table S 2). In this regard, the conclusions drawn from XPS measurements have to be mentioned again briefly. Electrolysis is associated with no variation in Fe^{3+} oxidation state, but with a probable dehydration of the surface end groups and with incorporation of phosphate anions which promotes the catalyst activity.

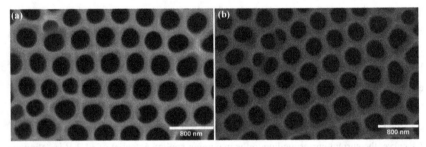

Figure 12. SEM micrographs of a nanostructured electrode (a) before and (b) after 16 hours of bulk electrolyis; $L = 14$ μm and $D = 332 \pm 29$ nm. The structures are unchanged as far as experimentally observable.

The electrochemical processes proceeding at the solid-liquid interface of the nanoporous electrode were investigated by the help of EIS, which was performed after the whole treatment. Figure 13 displays the Nyquist plot of the nanostructured electrode consisting of almost two semicircles gained for frequencies from 100 kHz to 20 mHz at +1.1 V. The data recorded at $\omega > 1$ can be assigned to slow reactions in the solid, whereas those recorded at $\omega > 1$ refer to solid-liquid interface processes. Since XPS measurements already revealed changes in the catalyst surface, the interface reactions are now in the focus of analysis and the fitting range is therefore reduced to $\omega \geq 0.20$ Hz (black data points). The Nyquist plot is now composed of a semicircle flanked by a straight line in the low frequency region indicating that the cells' impedance is controlled by charge transfer and diffusion[41] as opposed to pure charge transfer for planar electrodes. Furthermore, the range of impedances is reduced by about two orders of magnitude with respect to those of planar samples presented in Figure 10. However, the catalytic surface area was also enhanced by a factor of 20 compared to planar electrodes discussed in 5.2.1. Best fits (black curve in Figure 13) are achieved by employing the equivalent circuit with corresponding fitting parameters listed in Table 5. The calculated values of R_s agree with internal resistances between 300 and 500 Ω. R_{ct} is now in the range of kΩ compared to MΩ for planar samples which is expected for the enlargement of catalytic active sites. This model, also called Randles circuit,[43] describes mixed control systems and differs somewhat from the one created for planar substrates. The main distinction is the application of a finite Warburg impedance element W, which is, a constant phase element with $\alpha = 0.5$. W is added to R_{ct} to consider ion diffusion within the pores of finite length and thus regarding diffusion control.[44-46] Thus, diffusion of charges inside porous Fe_2O_3 electrodes is reflected in Nyquist plots. Under consideration of the values gained for planar electrodes (see Table 4), the surface expansion yields a reduction in R_{ct} by a factor of about 100, which is in the same order of magnitude as the surface area enhancement.

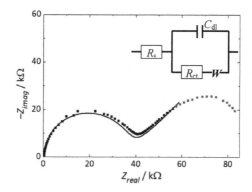

Figure 13. Nyquist plot for EIS data measured at +1.1 V from 100 kHz to 20 mHz for the electrochemically treated Fe_2O_3 electrode shown in Figure 12 (black + grey points); fit (black curve) is created for the fitting range of 100 kHz to 0.2 Hz (black points only). Additionally, equivalent Randles circuit used to conceptualize Fe_2O_3-liquid interfaces of nanostructured electrodes is presented; R_s = solution resistance, R_{ct} = charge transfer resistance, C_{dl} = double layer capacitor and W = finite Warburg impedance element; $A_{macro} = 3.1 \cdot 10^{-2}$ cm^2, $L = 14$ µm and $D = 332 \pm 29$ nm.

Table 5. Fitting parameters for the EIS data visualized in Figure 13 for the electrochemically treated Fe_2O_3 electrode; R_s = solution resistance, R_{ct} = charge transfer resistance, C_{dl} = double layer capacitor and W = finite Warburg impedance element.

circuit element	R_s / Ω	C_{dl} / nF	W / µF s$^{-1/2}$	R_{ct} / kΩ
	334 ± 2	75.4 ± 0.4	26.5 ± 0.5	35.7 ± 0.2

In summary, it can be concluded that electrochemically pre-treated nanoporous Fe_2O_3 electrodes were stable upon prolonged water oxidation at $\eta = 0.49$ V. Cyclic voltammograms further illustrate the current increase for pre-treatment and bulk electrolysis in comparison to the freshly prepared sample. Electrolysis is accompanied by dehydration of the catalyst (as well as a minor incorporation or adsorption of phosphate anions). This change causes an increase in surface roughness, although the overall morphology of the nanostructured samples is retained. The enhanced roughness results in an increased surface area and a concomitant increase in overall electrocatalytic activity. The expected competition between charge transfer and diffusion limitation for porous electrodes and the enhancement in catalytic active sites becomes evident from the analysis of EIS measurements.

5.3   Modification of nanoporous Fe_2O_3 electrodes

Before we move on to optimize the geometry of the electrode pores, let us ensure the quality of the electrical contact and of the material itself. We will perform this by investigating the effect of additional conductive platinum or aluminum doped zinc oxide (AZO) layers and annealing treatments on the water oxidation efficiency.

5.3.1 Annealed catalyst surfaces

The first modification approach is post-deposition annealing of Fe_2O_3 electrodes based on bilateral opened templates with a nickel backside contact. For this purpose, one membrane ($D = 222 \pm 13$ nm, $L = 22$ μm) was divided in halves, whereat one part was annealed under O_2 at 400 °C over 4 h in order to allow for direct comparison. These two parts were then used for the preparation of 3 to 4 electrodes. The same was performed for a second Fe_2O_3 membrane ($D = 291 \pm 17$ nm, $L = 7$ μm) in N_2 as annealing atmosphere. Since the templates differ in pore length and diameter, it is not possible to make a direct comparison. On the microscopic scale, annealed and untreated electrodes are not distinguishable as can be seen from SEM micrographs (see Figure S 3).

In the next step, the series of electrochemical experiments (1) to (5) from 5.2 was performed. The samples were pre-treated at +1.3 V for 1 h before current densities were gathered for bulk electrolysis. Effective current densities J and the corresponding logarithmic values $\log(J)$ gained for the investigated samples are listed in Table 6 and Table 7. Starting with the influence of O_2 annealing on the overall performance, an increase in J is apparent. Based on the calculated logarithmic values, an enhancement by half a magnitude is determined more precisely. This result can be exceeded by changing the atmosphere to N_2. Annealing now leads to an increased efficiency by almost one order of magnitude according to $\log(J)$ values. The depiction of $\log(J)$ values in a bar graph (see Figure 14) further illustrates the efficiency increase gained for annealing treatments.

Table 6. Average effective current densities J and corresponding logarithmic values log(J) measured for untreated and electrodes annealed under O_2 after 16 h of bulk electrolysis at +1.1 V. Each mean value was calculated from at least three measured values; $A_{macro} = 7.9 \cdot 10^{-3}$ cm^2, $D = 291 \pm 17$ nm and $L = 23$ μm.

J / μA cm^{-2}		log(J / μA cm^{-2})	
as deposited	annealed O_2	as deposited	annealed O_2
8.59 ± 3.5	26.07 ± 4.6	0.91 ± 0.2	1.41 ± 0.1

Table 7. Average effective current densities J and corresponding logarithmic values log(J) measured for untreated and electrodes annealed under N_2 after 16 h of bulk electrolysis at +1.1 V. Each mean value was calculated from at least three measured values; $A_{macro} = 7.9 \cdot 10^{-3}$ cm^2, $D = 279 \pm 36$ nm and $L = 7$ μm.

J / μA cm^{-2}		log(J / μA cm^{-2})	
as deposited	annealed N_2	as deposited	annealed N_2
7.23 ± 0.6	57.70 ± 23.7	0.86 ± 0.04	1.74 ± 0.2

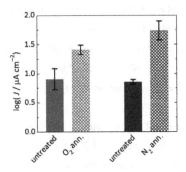

Figure 14. Logarithmic mean of the effective current densities measured at +1.1 V *vs.* Ag/AgCl for untreated and electrodes annealed under O_2 or N_2 as listed in Table 6 and Table 7. Comparison is drawn between untreated (light grey) and electrodes annealed in O_2 (light grey patterned) with $L = 23$ μm, and between untreated (grey) and electrodes annealed in N_2 (grey patterned) with $L = 7$ μm.

The changes occuring to the catalyst surface upon annealing under O_2 or N_2 can be studied by means of XPS at planar reference samples. For the overall comparison, annealed electrodes are compared with untreated samples and with annealed electrodes after 2 h of electrolysis at +1.3 V (see Figure 15, Table S 3 and Table S 4).

Starting with the consequences of annealing on the catalyst surface, a variation in the Fe 2p peak region is not observable (see Figure 15a), implying that ALD directly results in Fe(III) without any further changes in the surface composition. Regarding the O 1s peak region in Figure 15b and the recorded values for the Si 2p region (see Table S 3), annealed surfaces differ significantly from the ALD-prepared sample in the presence of peaks at ~ 532 eV and ~ 103 eV in the O 1s and Si 2p region, respectively, which are precisely attributable to SiO_2. It can therefore be concluded that annealing crystallizes the initially smooth, amorphous surface and causes a significant roughening of the layer which partly reveals SiO_2. This is in line with the knowledge that Fe_2O_3 ALD yields amorphous deposits at relatively low temperatures of 200 °C.[47]

Furthermore, the sample annealed in N_2 exhibits significant differences in the O 1s peak shape at 530.2 eV as well as a shift of 0.4 eV to higher binding energies with respect to the untreated sample. According to the literature, the O 1s peak of α-Fe_2O_3 is located at a binding energy of 529.8 eV, whereby the one for Fe_3O_4 is at 530.2 eV.[48] This is indeed the case for the ALD-prepared and in N_2 treated electrode, respectively. Thus, this shift can be ascribed to small quantities of Fe(II) in the catalyst surface.

Let us now consider the surface changes of annealed samples upon electrolysis. In the case of O_2 annealed electrodes, electrolysis is again associated with dehydration of the catalyst surface, as apparent from the increase in the Fe/O_{Fe} ratio (see Table 8). The surface roughness, however, is hardly altered, since the Fe/Si ratio and the SiO_2 peak at 532 eV do not significantly change with respect to the non-electrolyzed electrode (see Figure 15c). The electrochemical treatment, however, considerably alters the sample treated in N_2, since the peak related to SiO_2 at 532 eV and the Fe/Si ratio increases strongly (Figure 15d and Table 8). In this case, electrolysis is linked with further crystallization. This is indicative of a partly reduced iron oxide material, the morphology of which rearranges more profoundly upon subsequent electrochemical treatment. Finally after electrolysis, all three samples have, independently of their chemical history, similar peak shapes for the O 1s peak region (Figure 15e). Consequently, annealing in O_2 atmosphere yields rough and hydrated Fe_2O_3 surfaces, which are then dehydrated in electrolysis. The N_2 treatment, on the other hand, crystallizes the catalyst surface into a rough iron oxide consisting of small quantities of Fe(II), whereby the roughness of this surface further increases upon electrolysis.

Furthermore, phosphorous is not detectable for either annealed surfaces, indicating that phosphate anion incorporation does not proceed for these surfaces. Especially as annealing enhances the water splitting efficiency significantly, the contribution of PO_4^{3-} is probably less important than crystallization of the catalyst surface.

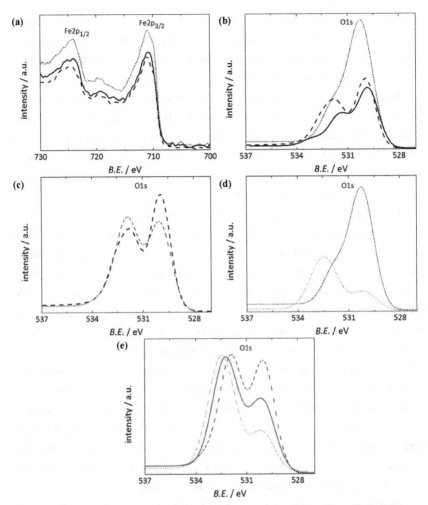

Figure 15. XPS spectra of the Fe 2p and O 1s peak regions for planar untreated, annealed and electrolyzed (2 h, +1.3 V) Fe_2O_3 electrodes. Comparison is drawn between untreated (black solid line), O_2 annealed (black dashed line) and N_2 annealed (black dotted line) electrodes in (a) and (b); between O_2 annealed electrodes without (black dashed line) and with subsequent electrolysis (grey dashed line) in (c); between N_2 annealed electrodes without (black dotted line) and with subsequent electrolysis (grey dotted line) in (d); and between untreated (grey solid line), O_2 annealed (grey dashed line) and N_2 annealed (grey dotted line) electrodes after electrolysis in (e).

Table 8. XPS element ratios of planar untreated and electrolyzed (2 h, +1.3 V) Fe_2O_3 electrodes; calculated from intensities listed in Table S 3 and Table S 4. The Fe/O_{Fe} ratio is computed from the total intensity of Fe and the intensity of the first and third O 1s peak except for the untreated sample, where only total intensities are used.

ratio	untreated	untreated + electrolysis	O_2 annealed	O_2 annealed + electrolysis	N_2 annealed	N_2 annealed + electrolysis
Fe/O_{Fe}	0.78	1.07	0.86	1.02	0.72	0.77
P/Fe	0	0.02	0	0	0	0
Si/Fe	0	0.25	0.10	0.14	0.05	0.68

Thus, surface area enhancement is the primary factor for the electrocatalytic efficiency, whereby N_2 atmosphere yields water oxidation turnover rates increased by one order of magnitude compared to unmodified electrodes. Consequently, post-deposition annealing of hematite electrodes prepared by ALD will stay in focus of this research as it is also commonly considered as necessary.[40, 49, 50]

5.3.2 Electrodes with underlying platinum layers

Besides the modification of the catalyst layer itself, pre-deposition of platinum as a conduction layer is tested. Metal layers were obtained by galvanic deposition for 10.5 min at alternating potential ($-400\,mV$ and $0\,V$) inside $22\,\mu m$ long Al_2O_3 templates which were previously opened unilateral and equipped with a $500 \pm 100\,nm$ thick electrical contact of copper and gold. The time of growth was selected in such a way that overgrowing was prevented and that the tube length accounted to approximately $10\,\mu m$. The successful growth is proven by the removal of templates in 1 M NaOH solution as depicted in Figure 16. SEM micrographs picture collapsed and broken platinum nanotubes grown on the electrical contact, whereby parts of Al_2O_3 are still visible (see Figure 16b). The length is indeed difficult to define. It can be assumed from the images that they are at least $5\,\mu m$ long, although according to a growth rate of approximately $1\,\mu m/min$, $10\,\mu m$ are more likely present. The wall thickness amounts to $\pm\,20\,nm$ for the alternating applied potential, as has been recently reported.[51] Accordingly, the diameter of the lower part of the pores is reduced by $\pm\,40\,nm$. The presence of platinum is further proven by EDX measurements (see Table S 5), accomplished after catalyst deposition ($\sim 10\,nm$ Fe_2O_3).

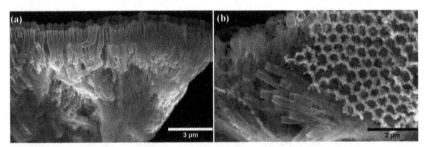

Figure 16. SEM micrographs of platinum nanotubes on a copper-gold contact including Al_2O_3 residues (visible in image b).

Table 9. Average effective current densities J and corresponding logarithmic values $\log(J)$ measured for Fe_2O_3 electrodes with and without platinum after 16 h of bulk electrolysis at $+1.1$ V. Each mean value was calculated from three measured values; $A_{macro} = 7.9 \cdot 10^{-3}\,cm^2$, $L = 22\,\mu m$, D(with Pt) $= 125 \pm 13\,nm$ and D(without Pt) $= 152 \pm 17\,nm$.

J / $\mu A\ cm^{-2}$		$\log(J$ / $\mu A\ cm^{-2})$	
Fe_2O_3	$Pt + Fe_2O_3$	Fe_2O_3	$Pt + Fe_2O_3$
19.13 ± 16.4	16.22 ± 8.7	1.18 ± 0.3	1.15 ± 0.3

In the next step, electrochemical measurements were performed and water splitting efficiency is compared for electrodes prepared of the Fe_2O_3 coated membranes with and without platinum layers as just described. All samples were pre-treated at +1.3 V for 1 h. In bulk electrolysis, the standard deviation in J obtained for several samples of each type is larger than the difference between the average values of the two types of samples. However, conductive platinum layers do not led to a performance increase based on the $log(J)$ values (see Table 9). The only change, which is observable, is a moderate resistance decrease from 1000 - 1200 Ω to 600 - 700 Ω[*].

Since the overall performance could not be improved sufficiently, the platinum layers might have been too short in order to overcome the semiconducting properties or charge transport within the solid is likely not the limiting factor at the relatively low current densities investigated here.

---

[*] For these electrodes higher internal resistances were measured compared to previously discussed electrodes. This is due to smaller coated pore diameters of 125 ± 13 nm and 152 ± 17 nm for electrodes with and without Pt layers, respectively.

5.3.3 Electrodes with underlying AZO layers

Instead of platinum, AZO is tested as a electrical conductor for the same templates, and samples are compared with the unmodified Fe_2O_3 electrodes introduced above. By means of ALD, approximately 10 nm of AZO were gained, whereat the process was built up by alternating cycles of ZnO and Al_2O_3 in the ratio 9:1. This ratio was chosen according to studies of Elam and George[52] on ALD of AZO of varying compositions. For more than 86% of ZnO cycles, they observed the best agreement between the ALD cycle ratio and the actually deposited AZO composition. In order to measure the presence of aluminum in the deposit, a piece of a titania tube was coated additionally and EDX measurements were subsequently performed for eight different points what from mean values were calculated (see Table 10). According to the weight percentages gained for zinc and aluminum, a ratio of 11:1 with a standard deviation of $\sigma = 1.5$ can be computed for Zn:Al. This ratio corresponds approximately to the intended ratio. AZO coated templates ($L = 22 \mu m$) with a corresponding pore diameter of 260 ± 15 nm are presented in Figure 17a. In the next step, these membranes obtained a Fe_2O_3 coating of ~ 10 nm thickness (see Figure 17b).

Table 10. Mean weight percentages (wt.%) calculated from EDX data for eight points on the AZO (~ 10 nm) coated titania tube.

element line	C K	O K	Al K	Ti K	Zn K
wt.%	6.42	11.15	0.34	78.35	3.73

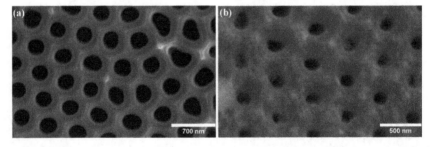

Figure 17. SEM micrographs of (a) the Al_2O_3 template coated with ~ 10 nm of AZO and (b) the same sample after Fe_2O_3 deposition. The micrographs demonstrate the reduction of pore diameter associated with the Fe_2O_3 coating.

Table 11. Average effective current densities J and corresponding logarithmic values log(J) measured for Fe_2O_3 electrodes with and without AZO after 16 h of bulk electrolysis at +1.1 V. Each mean value was calculated from at least two measured values; $A_{macro} = 7.9 \cdot 10^{-3}$ cm^2, $L = 22$ μm, D(with AZO) = 171 ± 18 nm and D(without AZO) = 152 ± 17 nm, $L = 22$ μm.

J / μA cm^{-2}		log(J / μA cm^{-2})	
Fe_2O_3	AZO + Fe_2O_3	Fe_2O_3	AZO + Fe_2O_3
19.13 ± 16.4	1.77 ± 0.6	1.18 ± 0.3	0.24 ± 0.1

The electrochemical activity was studied in comparison to unmodified electrodes. All samples were pre-treated at +1.3 V for 1 h before water oxidation efficiency was investigated. The modified electrodes yield current densities which are one magnitude smaller than those measured for unmodified Fe_2O_3 electrodes. In order to find an explanation for the performance diminution, the resistivity is determined by means of I-V curves recorded for Al_2O_3 templates coated with AZO. Such an I-V curve is presented in Figure 18. All samples reveal deviations from a linear relation between the applied voltage and measured current. This deviation from Ohmic behavior is due to the diode behavior of the contacts present in the system. A rough estimate of the specific volume resistance ρ is calculated for $E = +0.5$ V (see Table 12), since the curve shape is almost linear between 0 V and +0.5 V. ρ amounts approximately to $1.2 \cdot 10^5$ Ω cm, which is indeed eight magnitudes higher than resistivities reported for AZO films grown by ALD or magnetron sputtering.[53, 54] This high resistivity might be related to solid reactions at the Fe_2O_3-ZnO interface. Alternatively, zinc oxide has potentially been decomposed by protons which were generated during electrolysis.

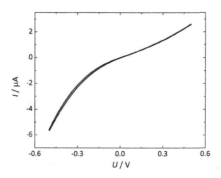

Figure 18. I-V curve of an electrode based on Al_2O_3 coated with ~ 10 nm of AZO; $L = 22$ μm, $D = 260 ± 15$ nm.

The importance of post-annealing is regarded as another relevant aspect. Post-deposition annealing especially for AZO films grown by RF magnetron sputtering is quite common.[55, 56] Surfaces were hereby partly annealed in-situ under argon atmosphere or in N_2 preferentially at 450 °C for 2 h. This treatment under N_2 atmosphere was chosen in order to study the influence on the resistivity. Unfortunately, ρ is even higher for annealed AZO surfaces (see Table 12). Thus, these surfaces are of little use for the enhancement of the water splitting efficiency.

Table 12. Calculated resistances for an applied potential of $E = +0.5$ V according to I-V curves of Al_2O_3 templates coated with ~ 10 nm of AZO for untreated and annealed samples (2 h at 450 °C under N_2); $L = 22$ μm, $D = 260 \pm 15$ nm, $A_{macro} = 3.1 \cdot 10^{-2}$ cm²; R = calculated resistance according to Ohm's law, A = effective area of porous system, ρ = specific volume resistance ($\rho = R \cdot \frac{A}{l}$).

sample	I / A	R / Ω	A / cm²	ρ / Ω cm
AZO untreated	$2.59 \cdot 10^{-6}$	$1.93 \cdot 10^5$	$1.93 \cdot 10^{-3}$	$1.2 \cdot 10^5$
AZO annealed	$3.66 \cdot 10^{-10}$	$1.37 \cdot 10^9$	$1.93 \cdot 10^{-3}$	$1.9 \cdot 10^8$

According to the performance diminution for Fe_2O_3 electrodes modified with AZO, a lot of parameters such as the ratio of Al to Zn, the growth conditions and a potential post-deposition annealing have to be optimized in order to apply AZO as underlying conductive layer. In this regard, the aspect of Zn etching by the $AlMe_3$ precursor during the Al_2O_3 ALD and the related deficits of ZnO as assumed by Elam et al. [52] should be considered.

5.4   Improvement of electrode performance by variation of pore length

Besides the modification of the electrode composition, the performance can be improved by tuning the Fe_2O_3 surface microstructure. Especially in this regard, a systematic investigation of how the electrocatalytic current varies with the specific surface area is of interest. For this purpose, several samples that feature almost the same diameter of $D \sim 300$ nm (see chapter 5.1) and interpore distance ($D_p = 500 \pm 30$ nm), but vary in pore length L from 7 to 23 µm were studied concerning their effective current densities gained for prolonged water electrolysis. All samples received 2 h of pre-treatment at +1.3 V followed by 16 h of bulk electrolysis at reduced η (0.49 V).

The elongation of pores from 7 to 14 µm results in an almost linear increase in the effective current density J (see Figure 19 and Table 13). Of special note is that the average current density of $J = 2.58 \cdot 10^{-1}$ µA cm^{-2} recorded for planar reference samples (black cross) perfectly fits into this correlation. For the 14 µm long pores a maximum is reached with $J = 17$ µA cm^{-2}. This value of J corresponds to a performance enhancement of almost two orders of magnitude with respect to planar electrodes (see log(J) values). Further extension of L up to 23 µm is correlated with a loss in effective current density. This reduction in the electrochemical activity is indicative of the limitation by mass and/or charge transport. Since the pore lengthening does not result in a simple plateau, it can further be assumed that the ohmic resistance of the semiconductor plays an additional role here.

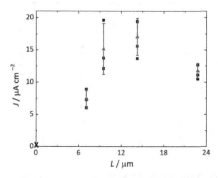

Figure 19. Stabilized electrocatalytic currents recorded at $E = +1.1$ V after 16 h for structured Fe_2O_3 electrodes with varying length L, compared with planar reference samples (black cross); $D \sim 300$ nm, average values and standard deviations σ for J are visualized in grey; black data points correspond to individual values. σ for L is ≤ 0.1 µm.

Table 13. Stabilized electrocatalytic currents J presented in Figure 19 and corresponding logarithmic values log(J) of structured Fe_2O_3 electrodes with varying length L; $D \sim 300$ nm.

L / µm	J / µA cm^{-2}	log(J / µA cm^{-2})
0	$2.58 \pm 1.4 \cdot 10^{-1}$	-0.59 ± 0.08
6.8 ± 0.3	7.42 ± 1.4	0.87 ± 0.08
9.5 ± 0.1	15.15 ± 3.9	1.18 ± 0.1
14.0 ± 0.3	17.03 ± 2.9	1.23 ± 0.07
22.8 ± 0.1	11.78 ± 1.1	1.07 ± 0.04

In order to connect this observation with already published results on Fe_2O_3 electrodes basing on AAO templates,[3] effective current densities for $\eta = 0.69$ V were gathered for all stable samples after additional 2 h at $E = +1.3$ V. The values are displayed in Figure 20. Although the standard deviation is quite high for each two electrodes, a similar linear increase in J is observable for the pore elongation. However, the maximum value is already reached for 10 µm pores. According to the higher applied overpotential, this is in line with the results made before. Furthermore, the comparison with data recorded by Gemmer *et al.*[3] entails that current densities reach the same magnitude. Nevertheless, an immediate comparison may not be drawn, since the majority of the samples was hardly stable at $\eta = 0.69$ V, probably due to the thinner nickel contact.

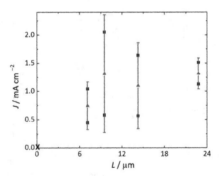

Figure 20. Stabilized electrocatalytic currents recorded at $E = +1.3$ V after additional 2 h for structured Fe_2O_3 electrodes with varying length L, compared with planar reference samples (black cross); $D \sim 300$ nm, average values and standard deviations σ for J are visualized in grey; black data points correspond to individual values. σ for L are ≤ 0.1 µm.

Our interpretation of the electrochemical behavior can be further investigated by means of EIS. Therefore, EIS was performed for all samples after bulk electrolysis. Representative Nyquist plots of a planar sample and two structured electrodes with 14 μm and 23 μm of pore length are shown in Figure 21. Values are plotted for frequencies ranging from 100 kHz to 0.2 Hz at +1.1 V. It is clearly visible that the shape of the Nyquist plot changes from a semicircle (see Figure 21a) for planar electrodes to a semicircle accompanied by a straight line (Figure 21b) for structured electrodes, towards an almost straight line (Figure 21c) for pores with $L = 23$ μm. This variation in curve shape can be attributed to a transition from charge transfer limited electrochemical processes indicated by the semicircle to diffusion limitation apparent as a straight line. Obviously, electrodes based on 14 μm long pores are still controlled by charge transfer and diffusion, whereas for pores of 23 μm length electrochemical processes are then limited by diffusion. Similar results on transitions from electron transfer kinetics to diffusion control can be found in the literature for CNTs of varying length.[41, 44]

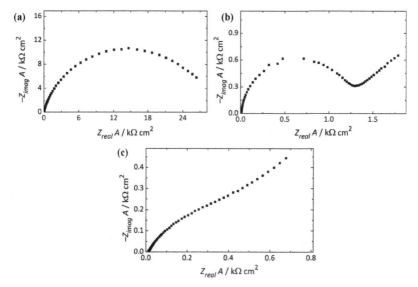

Figure 21. Nyquist plots for EIS data measured at +1.1 V from 100 kHz to 0.2 Hz for (a) a planar Fe_2O_3 sample, (b) a structured electrode of $L = 14$ μm and (c) a structured electrode of $L = 23$ μm after bulk electrolysis.

Furthermore, it should be mentioned that the charge transfer resistance R_{ct}, which is reflected by the semicircle diameter, reduces for increasing surface area as expected for the enlargement of catalytic active sites.

These EIS results confirm the previously expressed assumption of mass transport limitation for electrodes with pores of 23 μm length. Whether the accurate limit is at 14 μm or between 14 and 23 μm, will be a subject for further studies. Likewise, pore lengths of less than 7 μm down to 1 μm have to be investigated in order to complete the systematic study.

6   Conclusions and outlook

This work was focused on nanostructured Fe_2O_3 electrodes for electrochemical water oxidation at neutral pH and 0.49 V applied overpotential.

Fe_2O_3 has sufficient electrical conductivity at the relatively low current densities investigated here, so that additional conductive layers do not yield an improvement in the electrode performance.

XPS and EIS characterizations of the electrodes reveal that ALD of the catalyst results in a smooth and hydrated Fe_2O_3.

Subsequent prolonged electrolysis is accompanied by crystallization into rough and anhydrous Fe_2O_3 as proven by XPS. The decrease in charge transfer resistance, quantified by means of EIS, further confirmed that electrolysis leads to the crystallization of the amorphous surface associated with surface area enhancement.

Annealing treatments of the catalyst surfaces yield crystallized, rough and hydrated iron oxides for O_2 and N_2 atmospheres. Additionally, annealing in N_2 probably results in surfaces mainly consisting of Fe(III) and partly of Fe(II) species, respectively. This surface undergoes the most extensive morphology rearrangement upon subsequent electrochemical treatment.

For these annealing treatments, the water splitting efficiency is improved by almost one order of magnitude (in the case of N_2 annealing) with respect to untreated samples

Thus, the electrocatalytic activity is primarily related to the catalyst surface area. The relation between catalytic current and surface area is further proven by pore elongation. The surface area was increased until the point was reached, where the overall performance is no longer limited by charge transfer, but solely by mass transport.

The water splitting efficiency is thereby enhanced by two orders of magnitude compared to planar electrodes.

In this work, the strongest increase in water oxidation efficiency by 2.5 orders of magnitude with respect to planar electrodes is gained for a nanostructured electrode annealed in N_2 atmosphere.

By combining the ideal pore length with the annealing treatment in N_2 atmosphere, an optimized nanostructured Fe_2O_3 electrode is gained, which can be applied for photoelectrochemical water splitting in future.

References

[1]   A. J. Roberts, R. C. T. Slade, *ECS Trans.* **2010**, *28*, 33.

[2]   S. Chaudhari, D. Bhattacharjya, J.-S. Yu, *RSC Adv.* **2013**, *3*, 25120.

[3]   J. Gemmer, Y. Hinrichsen, A. Abel, J. Bachmann, *J. Catal.* **2012**, *290*, 220.

[4]   K. Sivula, F. Le Formal, M. Grätzel, *ChemSusChem* **2011**, *4*, 432.

[5]   P. Yang, Y. Ding, Z. Lin, Z. Chen, Y. Li, P. Qiang, M. Ebrahimi, W. Mai, C. P. Wong, Z. L. Wang, *Nano Lett.* **2014**, *14*, 731.

[6]   J. W. Morgan, E. Anders, *Proc. Natl. Acad. Sci. USA* **1980**, *77*, 6973.

[7]   N. Iordanova, M. Dupuis, K. M. Rosso, *J. Chem. Phys.* **2005**, *122*, 144305.

[8]   D. Cao, W. Luo, J. Feng, X. Zhao, Z. Li, Z. Zou, *Energy Environ. Sci.* **2014**, *7*, 752.

[9]   Z. Fu, T. Jiang, L. Zhang, B. Liu, D. Wang, L. Wang, T. Xie, *J. Mater. Chem. A* **2014**, *2*, 13705.

[10]   A. Kay, I. Cesar, M. Grätzel, *J. Am. Chem. Soc.* **2006**, *128*, 15714.

[11]   C.-Y. Lee, L. Wang, Y. Kado, R. Kirchgeorg, P. Schmuki, *Electrochem. Commun.* **2013**, *34*, 308.

[12]   Y. Liu, Y.-X. Yu, W.-D. Zhang, *Electrochim. Acta* **2012**, *59*, 121.

[13]   Y. Lin, G. Yuan, S. Sheehan, S. Zhou, D. Wan, *Energy Environ. Sci.* **2011**, *4*, 4862.

[14]   U. Bjoerksten, J. Moser, M. Grätzel, *Chem. Mater.* **1994**, *6*, 858.

[15]   N. T. Hahn, H. Ye, D. W. Flaherty, A. J. Bard, C. B. Mullins, *ACS Nano* **2010**, *4*, 1977.

[16]   M. Marelli, A. Naldoni, A. Minguzzi, M. Allieta, T. Virgili, G. Scavia, S. Recchia, R. Psaro, V. Dal Santo, *ACS Appl. Mater. Interfaces* **2014**, *6*, 11997.

[17]   H. Masuda, H. Tanaka, N. Baba, *Chem. Lett.* **1990**, *4*, 621.

[18]   J. Bachmann, R. Zierold, Y. T. Chong, R. Hauert, C. Sturm, R. Schmidt-Grund, B. Rheinländer, M. Grundmann, U. Gösele, K. Nielsch, *Angew. Chem. Int. Ed.* **2008**, *47*, 6177.

[19]   J. Bachmann, J. Escrig, K. Pitzschel, J. M. M. Moreno, J. Jing, D. Görlitz, D. Altbir, K. Nielsch, *J. Appl. Phys.* **2009**, *105*, 07B521.

[20]   M. Grätzel, *Nature* **2001**, *414*, 338.

[21]   A. Mao, N. Park, G. Han, J. Park, *Nanotechnology* **2011**, *22*, 175703.

[22]   W. Zhang, Y. Chen, *J. Nanopart. Res.* **2013**, *15*, 1334/1.

[23]   M. Thomalla, *Development of nano-structured injection solar cell with WS₂ absorber*, Dissertation, Freie Universität Berlin, **2007**.

[24] C. H. Hamann, W. Vielstich, *Elektrochemie*, Weinheim, WILEY-VHC Verlag, **2005**.

[25] J. K. Leland, A. J. Bard, *J. Phys. Chem.* **1987**, *91*, 5076.

[26] S. M. Ahmed, J. Leduc, S. F. Haller, *J. Phys. Chem.* **1988**, *92*, 6655.

[27] A. Fujishima, K. Honda, *Nature* **1972**, *238*, 37.

[28] B. Klahr, S. Gimenez, F. Fabregat-Santiago, T. Hamann, J. Bisquert, *J. Am. Chem. Soc.* **2012**, *134*, 4294.

[29] A. Lasia, in *Modern Aspects of Electrochemistry*, Bd. 32, (Herausgegeben von B. Conway, J. Bockris, R. White), Springer US, **2002**, 143.

[30] E. Barsoukov, J. R. Macdonald, *Impedance Spectroscopy, Theory, Experiment, and Applications*, Hoboken, John Wiley and Sons, **2005**.

[31] http://www.arradiance.com/Index_Files/pdf%20files/GemStar-6%20Data%20Sheet_2014.pdf#view=FitV. [2014-09-25]

[32] K. Nielsch, J. Choi, K. Schwirn, R. B. Wehrspohn, U. Gösele, *Nano Lett.* **2002**, *2*, 677.

[33] C. Wagner, W. Riggs, L. Davis, J. Moulder, G. Muilenberg, *Handbook of X-ray Photoelectron Spectroscopy*, Minnesota, Perkin-Elmer Corporation Physical Electronics Division, **1978**.

[34] P.-z. Yang, L.-m. Liu, J.-h. Mo, W. Yang, *Semicond. Sci. Technol.* **2010**, *25*, 045017.

[35] M. Engelhard, C. Evans, T. A. Land, A. J. Nelson, *Surf. Sci. Spectra* **2001**, *8*, 56.

[36] Y. Surendranath, M. W. Kanan, D. G. Nocera, *J. Am. Chem. Soc.* **2010**, *132*, 16501.

[37] M. W. Kanan, D. G. Nocera, *Science* **2008**, *321*, 1072.

[38] J.-B. Jorcin, M. E. Orazem, N. Pébère, B. Tribollet, *Electrochim. Acta* **2006**, *51*, 1473.

[39] U. Rammelt, G. Reinhard, *Electrochim. Acta* **1990**, *35*, 1045.

[40] J. Engel, H. L. Tuller, *Phys. Chem. Chem. Phys.* **2014**, *16*, 11374.

[41] S. Siddiqui, P. U. Arumugam, H. Chen, J. Li, M. Meyyappan, *ACS Nano* **2010**, *4*, 955.

[42] K. K. Lee, S. Deng, H. M. Fan, S. Mhaisalkar, H. R. Tan, E. S. Tok, K. P. Loh, W. S. Chin, C. H. Sow, *Nanoscale* **2012**, *4*, 2958.

[43] J. E. B. Randles, *Discuss. Faraday Soc.* **1947**, *1*, 11.

[44] M. Pacios, I. Martín-Fernández, R. Villa, P. Godignon, M. D. Valle, J. Bartrolí, M. Esplandiu, *Carbon Nanotubes as Suitable Electrochemical Platforms for Metalloprotein Sensors and Genosensors*, Carbon Nanotubes - Growth and Applications, **2011**.

[45] E. Warburg, *Ann. Physk.* **1899**, *67*, 493.

[46] F. Scholz, *Electroanalytical Methods: Guide to Experiments and Applications*, Berlin-Heidelberg, Springer Verlag, **2010**.

[47] A. B. F. Martinson, M. J. DeVries, J. A. Libera, S. T. Christensen, J. T. Hupp, M. J. Pellin, J. W. Elam, *J. Phys. Chem. C* **2011**, *115*, 4333.

[48] R. Cornell, U. Schwertmann, *The iron oxides - Structure, properties, reactions, occurences and uses*, Weinheim, WILEY-VHC Verlag, **2003**.

[49] B. M. Klahr, A. B. F. Martinson, T. W. Hamann, *Langmuir* **2010**, *27*, 461.

[50] G. Wang, Y. Ling, D. A. Wheeler, K. E. N. George, K. Horsley, C. Heske, J. Z. Zhang, Y. Li, *Nano Lett.* **2011**, *11*, 3503.

[51] V. Roscher, M. Licklederer, J. Schumacher, G. Reyes Rios, B. Hoffmann, S. Christiansen, J. Bachmann, *Dalton Trans.* **2014**, *43*, 4345.

[52] J. W. Elam, S. M. George, *Chem. Mater.* **2003**, *15*, 1020.

[53] J. Lee, D. Lee, D. Lim, K. Yang, *Thin Solid Films* **2007**, *515*, 6094.

[54] G. Luka, B. Witkowski, L. Wachnicki, R. Jakiela, I. Virt, M. Andrzejczuk, M. Lewandowska, M. Godlewski, *Mater. Sci. Eng., B* **2014**, *186*, 15.

[55] N.-F. Shih, J.-Z. Chen, Y.-L. Jiang, *Adv. Mat. Sci. Eng.* **2013**, 1.

[56] D.-K. Kim, C.-B. Park, *J. Mater. Sci.: Mater. Electron.* **2014**, *25*, 1589.

Supporting information

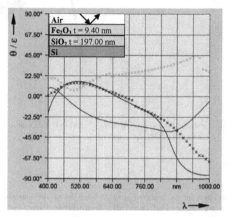

Figure S 1. Determination of the deposited Fe_2O_3 layer thickness by fitting the ellipsometric data (crosses), the orientation Θ and ellipticity ε, to an optical model (solid line).

Table S 1. Results of microscopic area calculation and linear regression for electrodes of varying length; $A_{macro} = 0.03$ cm^2; L = pore length, A_{micro} = microscopic area, R^2 = coefficient of determination.

L / μm	A_{micro} / cm^2
6.8	0.92 ± 0.1
9.5	1.16 ± 0.1
14.0	2.28 ± 0.2
22.8	3.59 ± 0.2
$R^2 = 0.9709$	

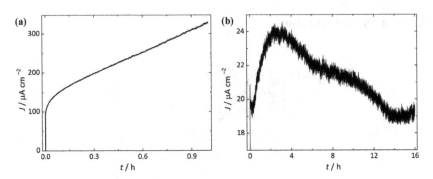

Figure S 2. Current density J recorded at (a) +1.3 V over 1 h and (b) +1.1 V at a structured Fe_2O_3 electrode; $A_{macro} = 3.1 \cdot 10^{-2} \ cm^2$, $L = 14 \ \mu m$ and $D = 332 \pm 29 \ nm$.

Table S 2. EDX data obtained for an analyzed area of $6 \times 6 \ \mu m$ of the nanoporous electrode pictured in Figure 12a and b: intensities are measured before and after electrochemical treatment.

element line	C K	O K	Al K	P K	Fe K	Ni K
intensity(before electrolysis) / a.u.	0.5	44.4	126.0	3.4	14.7	0.7
intensity(after electrolysis) / a.u.	0.5	42.6	128.9	3.5	13.3	0.6

Table S 3. Calculated binding energies *B.E.* and intensities *A* from the XPS high resolution scans for planar Fe_2O_3 electrodes as deposited and annealed under O_2 or N_2. P 2p peaks are not observable.

element	Peak	untreated		O_2 annealed		N_2 annealed	
		B.E. / eV	*A* / a.u.	*B.E.* / eV	*A* / a.u.	*B.E.* / eV	*A* / a.u.
Si 2p	1	-	-	100.06	32.50	101.76	160.56
	2	-	-	101.86	81.18	-	-
	3	-	-	102.76	116.01	-	-
P 2p	1	-	-	-	-	-	-
C 1s	1	284.62	993.26	284.72	644.91	284.63	575.41
	2	286.20	134.50	286.52	82.80	286.63	73.67
	3	288.39	78.57	288.54	82.73	288.85	42.30
N 1s	1	-	-	-	-	401.20	52.65
O 1s	1	529.82	1687.52	529.96	1819.33	530.22	3905.96
	2	531.46	891.91	531.69	1061.00	531.79	1207.46
	3	533.02	214.78	532.63	756.19	-	-
Fe 2p	1	710.75	1481.83	710.86	1546.78	710.68	1955.93
	2	724.29	708.55	724.59	660.70	724.48	860.37

Table S 4. Calculated binding energies *B.E.* and intensities *A* from the XPS high resolution scans for planar Fe_2O_3 electrolyzed electrodes (2 h, +1.3 V) as deposited and annealed under O_2 or N_2. N 1s peaks are not observable.

element	Peak	untreated + electrolysis		O_2 annealed + electrolysis		N_2 annealed + electrolysis	
		B.E. / eV	*A* / a.u.	*B.E.* / eV	*A* / a.u.	*B.E.* / eV	*A* / a.u.
Si 2p	1	101.98	198.35	102.07	262.45	99.55	13.35
	2	103.31	116.71	-	-	101.84	210.82
	3	-	-	-	-	103.41	122.32
P 2p	1	133.28	21.68	-	-	-	-
	2	139	6.03	-	-	-	-
C 1s	1	284.67	1135.78	284.55	765.59	284.78	1288.07
	2	286.37	97.05	284.55	111.33	286.34	286.34
	3	287.97	27.92	287.17	41.11	287.59	35.48
	4	-	-	288.68	57.20	288.95	55.88
N 1s	1	-	-	-	-	-	-
O 1s	1	530.09	1195.66	530.01	1618.54	530.10	663.60
	2	532.24	1849.62	531.91	1634.06	532.46	1990.11
	3	-	-	533.30	168.31	-	-
Fe 2p	1	710.91	852.05	710.95	1247.43	710.97	371.32
	2	724.59	422.40	724.71	570.80	724.47	136.59

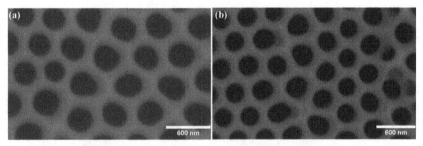

Figure S 3. SEM micrographs of (a) the untreated Fe_2O_3 electrode and (b) the same electrode annealed under N_2; $L = 7$ μm and $D = 291 \pm 17$ nm.

Table S 5. EDX data of a Fe_2O_3 electrode with an underlying platinum layer; $L = 22$ μm and $D = 152 \pm 17$ nm.

element line	C K	O K	Al K	Fe K	Cu K	Pt L	Au L
intensity / a.u.	1.5	56.6	299.8	15.9	2.6	1.1	0.5

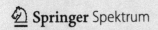